SCIENTIFIC AMERICAN | EDUCATIONAL PUBLISHING

SCIENTIFIC AMERICAN INVESTIGATES CAREERS IN SCIENCE

PALEONTOLOGIST

BY ROSIE BANKS

Published in 2025 by The Rosen Publishing Group
in association with Scientific American Educational Publishing
2544 Clinton Street, Buffalo NY 14224

Library of Congress Cataloging-in-Publication Data
Names: Banks, Rosie, 1978- author.
Title: Paleontologist / Rosie Banks.
Description: New York : Scientific American Educational Publishing, an imprint of Rosen Publishing, [2025] | Series: Scientific American investigates careers in science | Includes index. | Audience term: Juvenile
Identifiers: LCCN 2024038444 (print) | LCCN 2024038445 (ebook) | ISBN 9781725352414 library binding | ISBN 9781725352407 paperback | ISBN 9781725352421 ebook
Subjects: LCSH: Paleontologists–Vocational guidance–Juvenile literature. | Paleontology–Juvenile literature.
Classification: LCC QE714.7 .B273 2025 (print) | LCC QE714.7 (ebook) | DDC 560.23–dc23/eng/20240822
LC record available at https://lccn.loc.gov/2024038444
LC ebook record available at https://lccn.loc.gov/2024038445

Designer: Leslie Taylor
Editor: Therese Shea

Portions of this work were originally authored by Lori Dowell and published as *Be a Paleontolologist*. All new material in this edition is authored by Rosie Banks.

Photo credits: Cover (main) benedek/iStockphoto.com; series art (background) jijomathaidesigners/Shutterstock.com; p. 4 Wlad74/Shutterstock.com; p. 5 (top) Asia_K/Shutterstock.com, (bottom) RobertKacpura/Shutterstock.com; p. 7 paleontologist natural/Shutterstock.com; p. 9 (background) Biodiversity Heritage Library/Flickr.com, (inset Kuo) Hans A. Rosbach/File:Beijing Ancient Observatory 20090715-19 (cropped).jpg_commons.wikimedia.org, (inset Cuvier) Christies/File:Georges Cuvier 3.jpg_commons.wikimedia.org; p. 11 Nuntiya/Shutterstock.com; p. 12 Gricharduk/ File:Head on view of the statue of Mary Anning.jpg_commons.wikimedia.org; p. 13 (top) Jaroslav Moravcik/Shutterstock.com, (inset skull and torso) Everard Home (1756 - 1832)/File:AnningIchthyosaurSkull.jpg_commons.wikimedia.org, (inset trex) Puwadol Jaturawutthichai/Shutterstock.com; p. 15 DatraxMada/File:River with dinosaur footprints.png_commons.wikimedia.org; p. 17 LuFeeTheBear/iStockphoto.com; p. 18 Maris Grunskis/Shutterstock.com; p. 19 Luc V. de Zeeuw/Shutterstock.com; p. 21 (*Lokiceratops*) Ddinodano/File:Lokiceratops rangiformis.png_commons.wikimedia.org, (*Musankwa*) Levi Martinez-Reza/File:Musankwa LMR.png_commons_wikimedia.org, (*Natovenator*) Sungjin Lee, Yuong-Nam Lee, Philip J. Currie, Robin Sissons, Jin-Young Park, Su-Hwan Kim, Rinchen Barsbold & Khishigjav Tsogtbaatar/File:Natovenator hunting fish.png_commons.wikimedia.org, (*Gorgosaurus*) kamomeen/Shutterstock.com; p. 23 James Jiao/Shutterstock.com; p. 24 Featureflash Photo Agency/Shutterstock.com; p. 25 milehightraveler/iStockphoto.com; p. 27 Max Acronym/Shutterstock.com; p. 29 (main) Rattana/Shutterstock.com, (inset) Daniel Eskridge/Shutterstock.com.

Printed in the United States of America

CPSIA compliance information: Batch #CWSA25. For Further Information contact Rosen Publishing at 1-800-237-9932.

CONTENTS

Words in the glossary appear in **bold** type the first time they are used in the text.

THE SCIENCE OF PALEONTOLOGY

You've likely grown up knowing about dinosaurs and other ancient creatures. That knowledge was revealed by fossils. Scientists called paleontologists study the history of life based on fossil evidence, or proof.

Fossils are the remains, or traces of remains, of life. Most fossils are preserved by layers of **sediment** that covered and buried them. The soft parts of the body break down, leaving hard parts, such as bones and teeth. Water containing **minerals** can fill the tiny spaces in these parts. When the water dries up and the minerals harden, a rock-like object shaped like the body part is left. Fossils reveal a lot about life on ancient Earth.

Paleontologists don't just look for dinosaur fossils. Plant fossils also offer information about prehistoric life, or life before recorded history.

Fossils can form in different ways, but conditions have to be just right for remains to be preserved for so long.

How Fossils Can Form

In Sedimentary Rock (most fossils)

- Minerals replace plant and animal matter within sedimentary rock.
- A buried organism leaves behind an empty space called a cast. If the empty space fills with minerals that harden, it's called a mold.

Other Places (rare)

- Tree sap called amber can preserve whole organisms, including soft tissues.
- A quick burial in ice, tar pits, and other places can preserve tissue.

IN THE FIELD AND AT THE LAB

When you're at the beach, do you look for shells? Did you ever dig in dirt, just to see what you'd find? Both of those activities are a good preparation for becoming a paleontologist!

As part of their work, paleontologists carefully dig through dirt and rocks, looking for pieces of fossilized bones or traces. This is called fieldwork. Paleontologists also work in labs. They study fossils closely under **microscopes** and perform tests on them with chemicals. They have methods of discovering how old fossils are, the kind of climate they lived in, and much more about them!

FUN FACT

BODY FOSSILS ARE FROM THE BODY, SUCH AS BONES, TEETH, AND WASTE. TRACE FOSSILS ARE MARKS LEFT BEHIND, INCLUDING FOOTPRINTS OR IMPRESSIONS OF SKIN, FEATHERS, AND LEAVES.

Paleontologists sometimes must put together many fossils to get an idea of a whole form. This is a bit like putting together a puzzle!

ANCIENT PALEONTOLOGISTS

People have unearthed fossils for thousands of years on purpose and by accident. When they found fossils of unknown animals and plants, they made guesses about their origins. Around 540 BCE, ancient Greek thinker Xenophanes noticed fossils of seashells on dry land. He suggested that the land must have been underwater during an earlier period in Earth's history.

In the 11th century, Chinese scientist Shen Kuo proposed an early **theory** of climate change. He based his **hypothesis** on the discovery of fossilized bamboo in a place where the conditions weren't right for bamboo growth. He concluded the area was once cooler and wetter.

FUN FACT

MANY PALEONTOLOGISTS THINK DRAGONS IN OLD TALES WERE INSPIRED BY DINOSAUR BONE DISCOVERIES. PEOPLE WERE LIKELY TRYING TO MAKE SENSE OF SOMETHING THAT NO LONGER EXISTED IN THEIR TIME.

Shen Kuo

Georges Cuvier convinced other scientists that animals can and have died out, or become extinct.

Georges Cuvier

In the 1700s, fossil collecting became popular. Scientists began to group kinds of fossils and record details about them. French scientist Georges Cuvier (1769–1832) is called the father of paleontology. He compared animal fossils to still-living animals. He also had a theory that the farther down in rock that fossils were found, the older they were.

KINDS OF PALEONTOLOGISTS

People often think of fossils as big bones, but there are so many kinds, from ones you can't see with just your eyes to the largest dinosaur remains. Fossils can be a few thousand years old to 3.5 billion years old. Because of this wide range of time and subjects, paleontologists usually choose an area to specialize in, or focus on.

Micropaleontologists study microfossils, which are fossils that can only be seen with a microscope. Paleobotanists are paleontologists who study plant fossils such as wood, leaves, flowers, and seeds. Animal paleontologists often choose to be either invertebrate paleontologists, who study animals without a backbone, or vertebrate paleontologists, who study animals with backbones.

FUN FACT

INVERTEBRATE PALEONTOLOGISTS STUDIED SOME INVERTEBRATE MARINE FOSSILS FOUND IN THE DESERT OF NEVADA. THE 200-MILLION-YEAR-OLD FOSSILS SUGGEST THAT PARTS OF THAT DRY STATE WERE ONCE COVERED BY WATER!

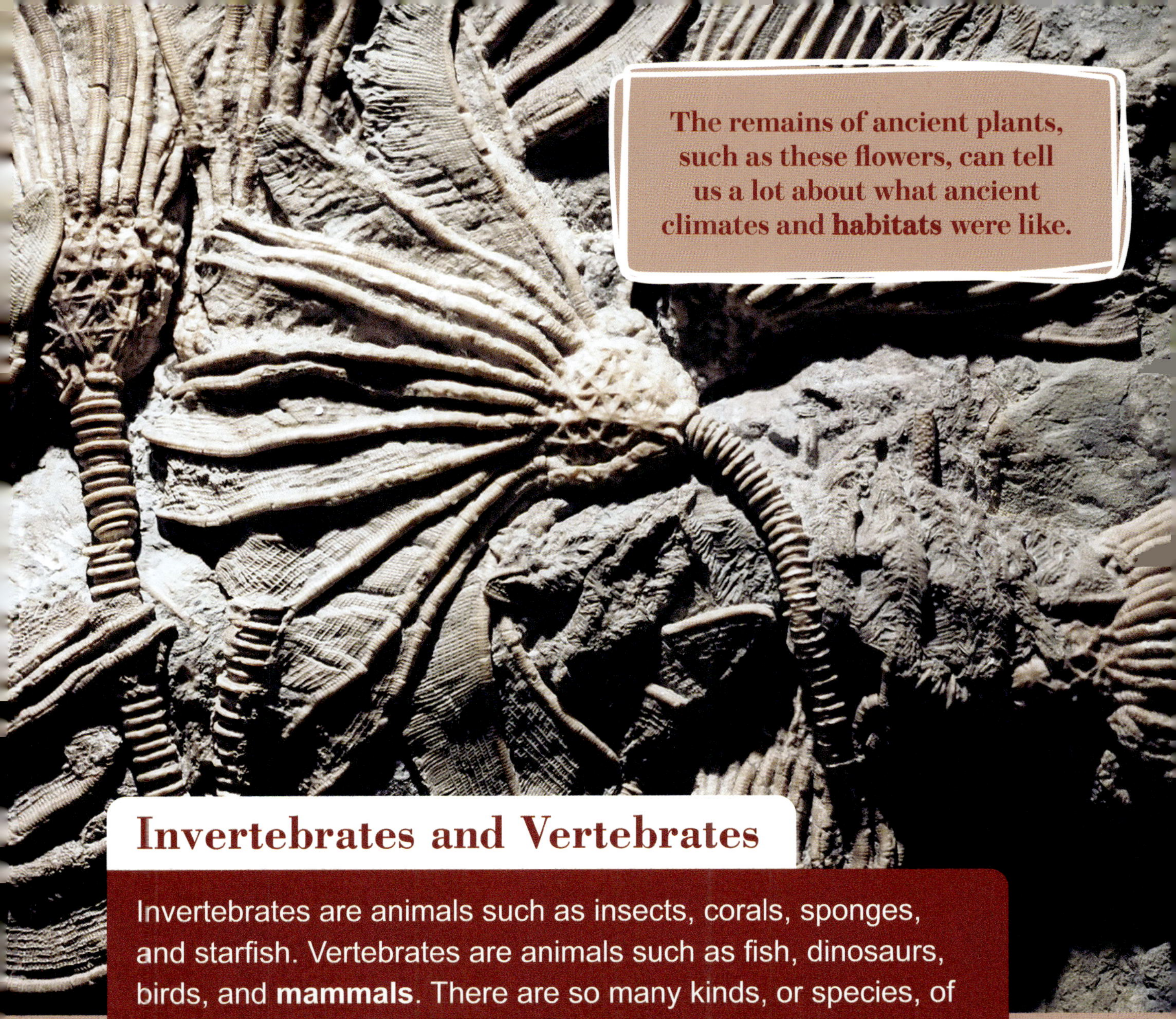

The remains of ancient plants, such as these flowers, can tell us a lot about what ancient climates and **habitats** were like.

Invertebrates and Vertebrates

Invertebrates are animals such as insects, corals, sponges, and starfish. Vertebrates are animals such as fish, dinosaurs, birds, and **mammals**. There are so many kinds, or species, of animals that it's easier to specialize in just one of these groups. However, vertebrate and invertebrate paleontologists may work together since different kinds of animal fossils are found near each other.

ANNING AND OTHERS

Mary Anning (1799–1847) was a fossil hunter in England whose findings helped the field of paleontology. She first began fossil hunting on the coast of the English Channel as a young girl. She sold the fossils to earn money for her family. When Mary was 12, she and her brother Joseph found parts of an *Ichthyosaurus*, an ancient marine reptile. She later found the fossils of a pterosaur, a flying reptile, and an ancient sharklike creature.

In the Gobi Desert of Mongolia, Roy Chapman Andrews (1884–1960) discovered the first fossilized dinosaur eggs in 1923, proving that some dinosaurs laid eggs. Andrews also found the fossils of the meat-eating *Velociraptor*.

Mary Anning

Joseph Anning found an *Ichthyosaurus* skull fossil, while Mary Anning found the torso, both pictured below, as well as the entire body of an *Ichthyosaurus*, above.

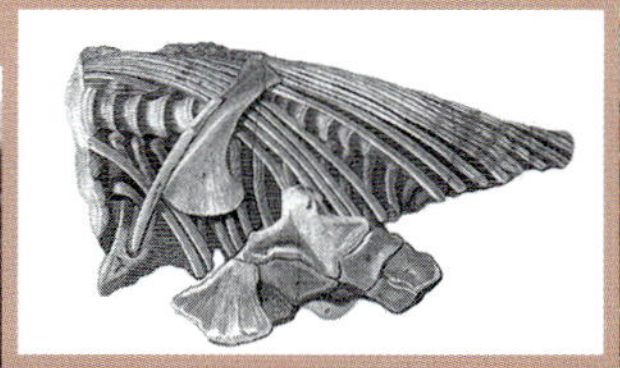

First *T. Rex* Fossils

Paleontologist Barnum Brown (1873–1963), nicknamed “Mr. Bones,” discovered the first *Tyrannosaurus rex* fossils, in 1902 in Montana. He discovered an almost complete *T. rex* six years later. The findings were shipped to the American Museum of Natural History in New York City. Large crowds gathered to see the remains of this “monster.”

American paleontologist Roland Bird (1899–1978) worked with Barnum Brown, collecting dinosaur fossils for the American Museum of Natural History. In 1938, he discovered a set of dinosaur footprints in a river in central Texas. These footprints, dating back 107 million years, are called Glen Rose Trackway, which is located in today's Dinosaur Valley State Park. Some of the footprints found in the Glen Rose Trackway measure 3 feet (0.9 m) long!

In the 1960s, paleontologist John H. Ostrom (1928–2005) suggested that birds **evolved** from dinosaurs. He compared fossils of an early bird to those of a birdlike dinosaur and offered the similarities as proof.

In 2022, very dry weather in Texas revealed 113-million-year-old dinosaur tracks in Dinosaur Valley State Park. The Paluxy River usually covers these tracks.

Extinction Ideas

In 1980, scientists Luis Alvarez and son Walter Alvarez found evidence an asteroid, a huge space rock, hit Earth about 65 million years ago. They suggested its impact caused tons of dust and other matter to be blown into the atmosphere, changing Earth's climate and causing a massive extinction event that killed off dinosaurs and other life-forms.

FUN FACT

MANY SCIENTISTS BELIEVE THE CRASH SITE OF THE ASTEROID THAT LED TO THE MASS EXTINCTION EVENT 65 MILLION YEARS AGO IS IN SOUTHERN MEXICO, PARTLY IN THE GULF OF MEXICO. IT'S CALLED THE CHICXULUB CRATER.

TOOLS OF PALEONTOLOGY

Sometimes, landslides or **earthquakes** uncover fossils. They may be found on the sides of rocky cliffs too. This makes a paleontologist's job easier, but they still need to use many different tools to dig out and dust off layers of earth and rock. They use large rock picks, small shovels called trowels, and drills to excavate, or unearth, fossils. Whenever hammering hard rock, they wear safety glasses and gloves to protect their eyes and hands. Paleontologists carry magnifying glasses to help them see smaller fossils.

Paleontologists also use **GPS** to keep track of where they are as well as discoveries' locations. They may use walkie-talkies to talk to each other too!

FUN FACT

DID YOU EVER USE A SIEVE AT THE BEACH? IT HAS SMALL HOLES THAT LET SAND FALL THROUGH, KEEPING STONES AND SHELLS IN THE SIEVE. PALEONTOLOGISTS USE THESE TOO.

Smaller picks and brushes are gentle ways to remove matter around fossils. Paleontologists don't want to destroy the fossils while trying to excavate them!

DATING FOSSILS

A fossil's age can be found using two main methods. One is called relative dating. Sedimentary rock forms as layers, each on top of the next. Bottom layers are older than those above. So, if a fossil is found in a bottom layer, it's older than fossils found above.

Absolute dating is a way of finding a more exact date. Certain elements in rock break down at a constant, or regular, rate. By measuring how much these elements have broken down, scientists can figure out how old something is. Sometimes it's easier to date the rocky matter around the fossil than the fossil itself.

FUN FACT

SCIENTISTS MAY BE ABLE TO TELL HOW OLD AN ANIMAL WAS WHEN IT DIED BY LOOKING AT THE DISTANCE BETWEEN THE GROWTH RINGS, OR MARKS, FOUND JUST BELOW THE SURFACE OF THE BONES.

Fossils have been found in the layers of sedimentary rock at Downpatrick Head in Ireland, shown here.

A NATIONAL MONUMENT

In a desert area in northeastern Utah and northwestern Colorado is Dinosaur National Monument. Paleontologist Earl Douglass discovered fossils there in 1909. He saw eight dinosaur tailbones sticking out of a hill. The remains date back 150 million years! It's thought that the dinosaurs didn't live and die in this area. River systems likely carried the remains there.

The area's so-called Wall of Bones has more than 1,500 dinosaur bones, including a *Stegosaurus* neck and an *Allosaurus* skull. These weren't all naturally exposed. Many paleontologists chipped away to show the fossils, so people could learn more about ancient creatures.

FUN FACT

A NATIONAL MONUMENT IS A PLACE THAT HAS GREAT BEAUTY OR HISTORIC OR SCIENTIFIC IMPORTANCE. DINOSAUR NATIONAL MONUMENT WAS ESTABLISHED BY PRESIDENT WOODROW WILSON IN 1915.

Many discoveries are made each year in paleontology. Each one helps us understand more about the prehistoric world.

More Dinosaur Discoveries

2019, new giant horned, plant-eating dinosaur called *Lokiceratops rangiformis* found in Montana

2022, first remains of swimming dinosaur *Natovenator* found in Mongolia

2023, largest penguin species ever on Earth discovered in New Zealand

2024, new dinosaur species called *Musankwa sanyatiensis* is found in Zimbabwe

2024, a new species of pterosaur is found on the Isle of Skye in Scotland

2023, remains of the dinosaur *Gorgosaurus* reveal what it had in its stomach before it died

WHAT TO STUDY

Paleontology combines a knowledge of earth science and biology. Paleontologists study different sciences in college, such as geology, biology, and chemistry.

Geology is the study of Earth, especially of rocks, soil, and minerals and the processes they undergo. Biology is the study of life and life processes. Since paleontology deals with forms of life trapped within Earth's geologic formations, you can understand why both these sciences would be important to a paleontologist's work. Studying the chemicals in bones and shells can help paleontologists learn about climates and habitats during the time the animal or plant was alive. This gives us a more complete picture of life back then.

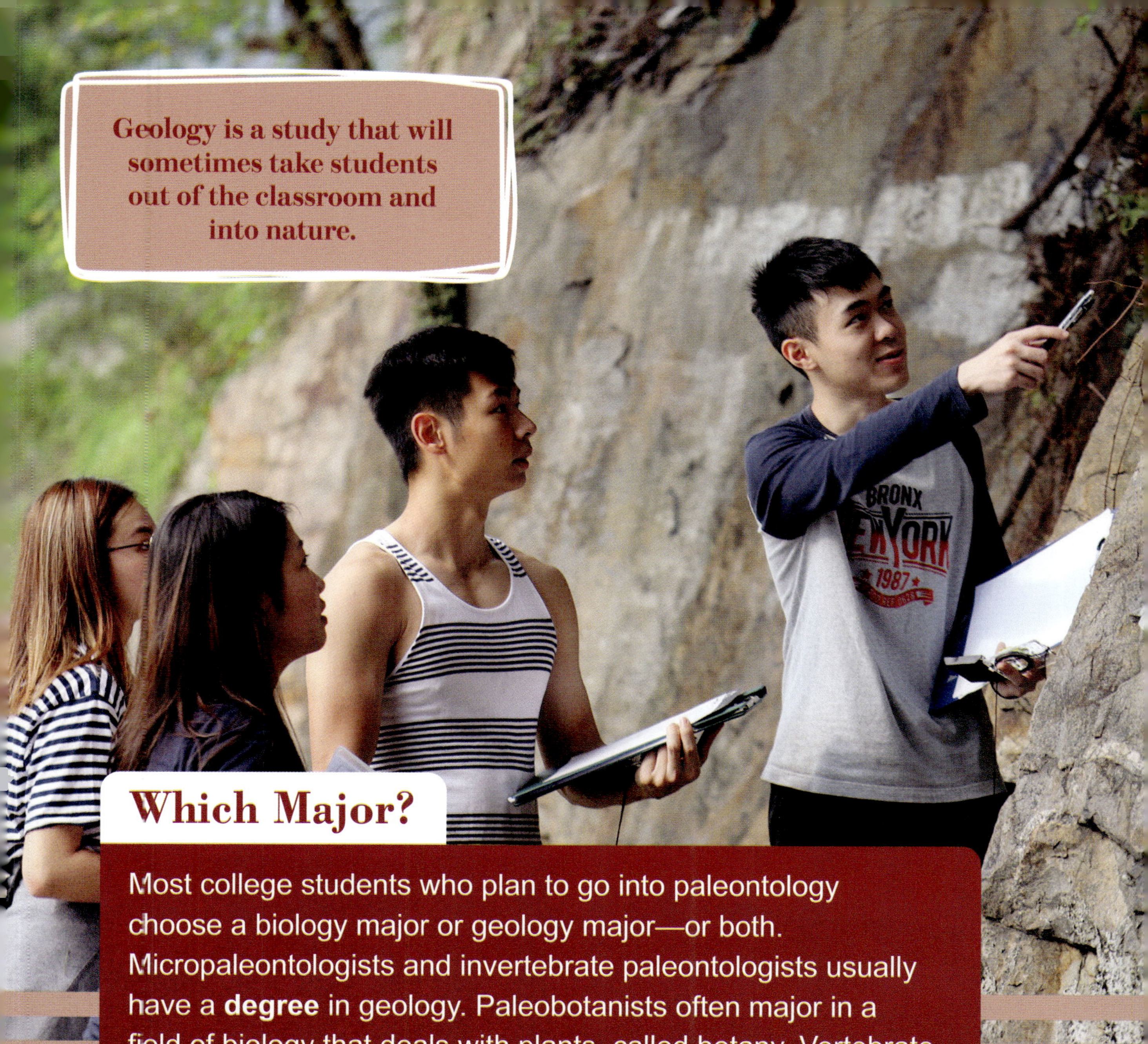

Geology is a study that will sometimes take students out of the classroom and into nature.

Which Major?

Most college students who plan to go into paleontology choose a biology major or geology major—or both. Micropaleontologists and invertebrate paleontologists usually have a **degree** in geology. Paleobotanists often major in a field of biology that deals with plants, called botany. Vertebrate paleontologists may focus on a field of biology called zoology, which is the study of animals.

Many careers in paleontology require a graduate degree. In graduate school, studies continue in the classroom and in labs, building on what students learned in college. Students will do more fieldwork too.

To receive a graduate degree, paleontology students complete **research** projects guided by a teacher. They write a long paper called a thesis explaining what they've discovered. After that, students choose their career path.

Some paleontologists work in museums or work on digs to collect fossils for museum display. Paleontologists are also employed by state or national governments. Still other paleontologists work for universities, teaching and doing research.

FUN FACT

PALEONTOLOGIST JACK HORNER WAS HIRED TO HELP WITH THE *JURASSIC PARK* AND *JURASSIC WORLD* MOVIES!

Paleontologists who work for governments may create maps of paleontological sites or identify fossils in state and national parks.

TECHNOLOGY TOO

Technology has added several tools to the field of paleontology. Machines called Geiger counters are used to locate elements underground that are often found in bones. Sonic technology sends sound waves into the ground. If the waves bounce back in a certain way, paleontologists know something might be buried there. CT (computed tomography) scans are a technology that uses **X-rays** to create an inside look at fossils, without harming them.

Computers and computer programs are also allowing paleontologists to record, share, and compare their discoveries and research more easily than ever.
A knowledge of computers would be helpful for a future paleontologist.

FUN FACT

PALEONTOLOGISTS HAVE USED CT SCANS TO CREATE MODELS OF BONES. THESE MODELS CAN BE USED BY COMPUTER PROGRAMS TO FIGURE OUT HOW PREHISTORIC CREATURES WALKED AND RAN!

Computers are a big part of lab work for the modern paleontologist.

GET STARTED NOW

There are only about 3,000 working paleontologists in the United States today. That means just a few paleontologists are experts in their specialized field. It can be hard to get a job in paleontology, but those who do love their work.

You don't have to wait until after college to start your paleontology career. Some museums organize digs for people interested in paleontology. This is a way to learn about fieldwork. The U.S. National Park Service also has a Junior Paleontologist Program. And don't forget that many young people have made paleontological discoveries by exploring parks, beaches, and other places in nature. You could be the next great paleontologist!

People interested in paleontology have curiosity and excitement about the world around them.

FUN FACT

IN 2020, A 12-YEAR-OLD NAMED NATHAN HRUSHKIN FOUND FOSSILS OF THE DINOSAUR *HADROSAUR* WHILE HIKING IN CANADA'S ALBERTAN BADLANDS.

GLOSSARY

degree: An official paper given to someone who has completed certain classes at a college or university.

earthquake: A shaking of the ground caused by the movement of Earth's crust.

evolve: To slowly change into a new, often more complex, form over thousands of years.

GPS: Stands for Global Positioning System. A system that uses satellite signals to locate places on Earth.

habitat: The natural place where an animal or plant lives.

hypothesis: An explanation based on known facts that explains an event or set of conditions.

mammal: A warm-blooded animal that has a backbone and hair, breathes air, and feeds milk to its young.

microscope: A tool used to view very small objects so they can be seen much larger and more clearly.

mineral: Matter in the ground that forms rocks.

research: Studying to find something new.

sediment: Matter, such as stones and sand, that is carried onto land or into the water by wind, water, or land movement. It forms sedimentary rock.

theory: An explanation based on facts that is generally accepted by scientists.

X-ray: A powerful type of energy that is similar to light but is invisible to the human eye.

FOR MORE INFORMATION

Books

Csotonyi, Julius. *Kid Paleontologist: Explore the Remarkable Dinosaurs, Fossils Finds, and Discoveries of the Prehistoric Era*. Nashville, TN: Applesauce Press, 2023.

Newland, Sonya. *Flying (Pterosaur) Reptiles: Classify the Features of Prehistoric Creatures*. London, UK: Franklin Watts, 2022.

Theodorou, Rod. *I Wonder Why Triceratops Had Horns: And Other Questions About Dinosaurs*. New York, NY: Kingfisher Books, 2024.

Websites

I Want to Be a Paleontologist! A Guide for Students
www.priweb.org/blog-post/i-want-to-be-a-paleontologist
Check out this career advice from the Paleontological Research Institution.

National Fossil Day
www.nps.gov/subjects/fossilday/kidsyouth.htm
Did you know there's a National Fossil Day? Learn more through National Park Service links.

Paleontology: The Big Dig
www.amnh.org/explore/ology/paleontology
The American Museum of Natural History presents a lot of interesting information about this field of science.

INDEX